DATE DUE

first starts

Snow and Ice

Joy Palmer

Austin, Texas

Editor: A. Patricia Sechi
Design: Shaun Barlow
Project Manager: Joyce Spicer
Electronic Production: Scott Melcer
Artwork: Kevin Wells
Cover Art: Kevin Wells
Picture Research: Ambreen Husain

Library of Congress Cataloging-in-Publication Data
Palmer, Joy.
Snow and ice / Joy Palmer.
p. cm. — (First starts)
Includes index.
Summary: An easy-to-read introduction to ice, snow, and other conditions which often accompany cold temperatures.
ISBN 0-8114-3414-1
1. Snow — Juvenile literature. 2. Ice — Juvenile literature. [1. Snow. 2. Ice.] I. Title. II. Series.
QC926.37.P35 1992
551.57'84—dc20 92-38438
CIP AC

Printed and bound in the United States

1 2 3 4 5 6 7 8 9 0 LB 98 97 96 95 94 93

Contents

What Are Snow and Ice?

Clouds are made of tiny droplets of water. When these freeze, they change into **ice crystals**. The crystals move around inside the cloud. They collide and stick together. When they are large enough and heavy enough, they fall out of the cloud. If the temperature is below freezing, they fall as snowflakes.

▽ As snowflakes fall on top of each other they pile up and cover the ground.

Sticking Flakes

As snowflakes fall from a cold cloud, they become softer and stick together very easily. Soft flakes can be rolled into balls and used to build interesting shapes. Every snowflake has six sides. Millions of them fall, and no two are ever the same. They are all slightly different. The clouds in the sky, and the snow that falls from them, are part of the world's **water cycle**.

▽ Snow is a good material to build with. These shapes have been built for a snow festival in Japan.

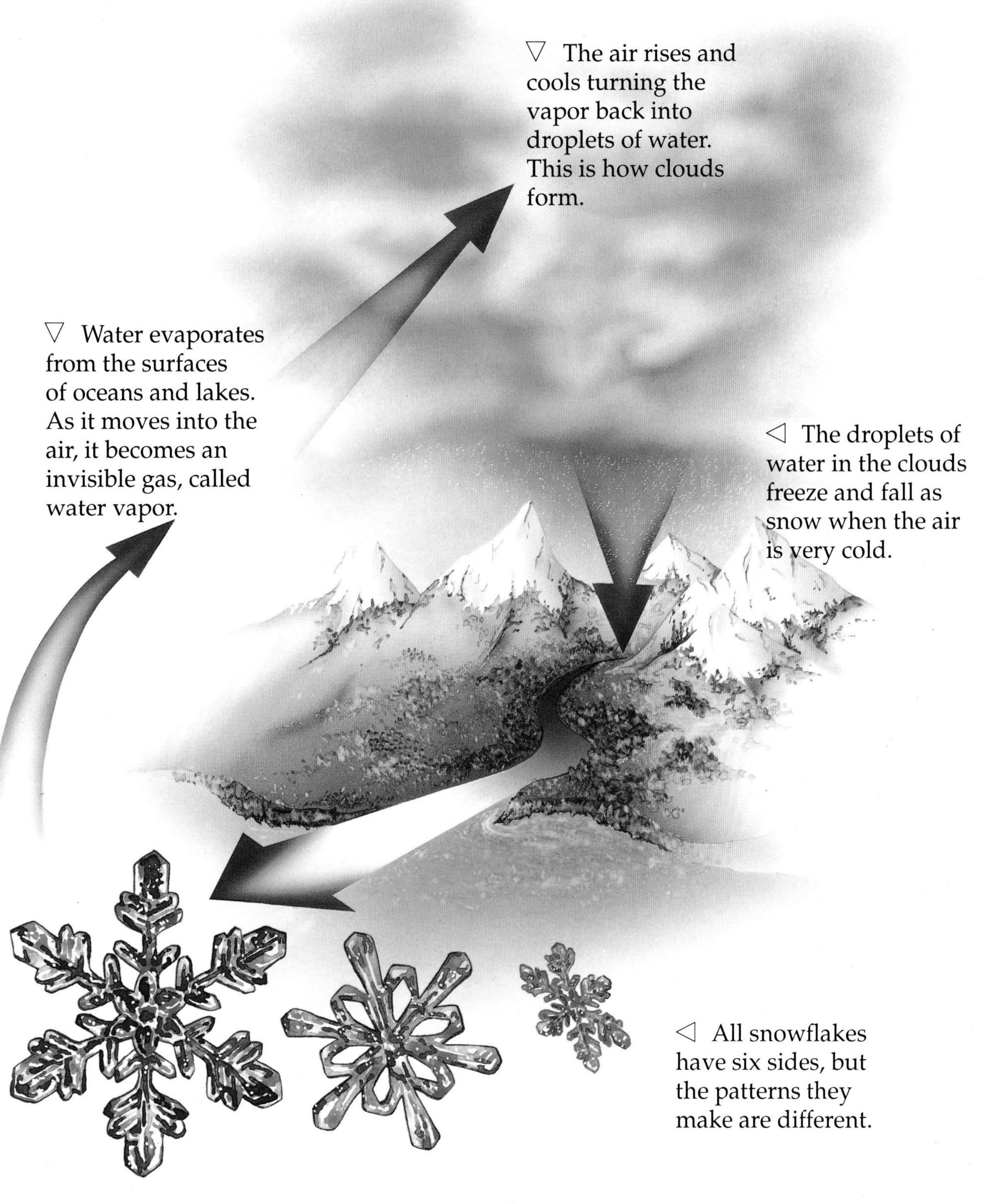

▽ The air rises and cools turning the vapor back into droplets of water. This is how clouds form.

▽ Water evaporates from the surfaces of oceans and lakes. As it moves into the air, it becomes an invisible gas, called water vapor.

◁ The droplets of water in the clouds freeze and fall as snow when the air is very cold.

◁ All snowflakes have six sides, but the patterns they make are different.

Ice from the Sky

Sometimes hard balls of ice form inside a storm cloud. They are tossed around inside the cloud and bump into water droplets. The droplets freeze onto the balls of ice, adding extra layers of ice. When the balls are heavy enough, they fall to the ground with great force and are called hailstones. Hailstones can cause a lot of damage.

▷ Hailstones fall to the ground with great force. They are usually about the size of a pea, but they can be as large as a melon!

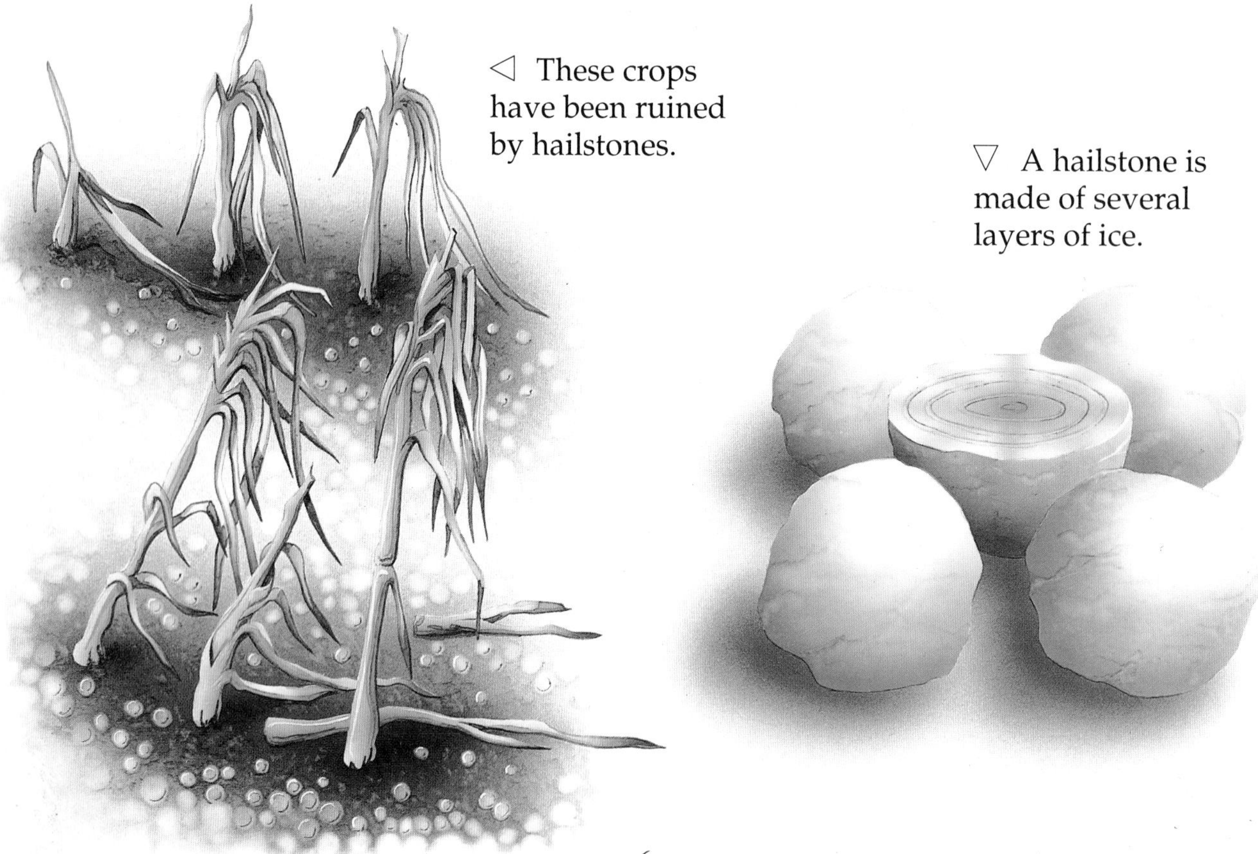

◁ These crops have been ruined by hailstones.

▽ A hailstone is made of several layers of ice.

Frost

During the night, when the sun does not shine, the ground cools down. In winter the **temperature** may fall below the **freezing point**. The water vapor in the night air will freeze when it meets something very cold, such as a blade of grass or a window. White ice crystals form. We call this frost.

▽ Trees and plants look all white the morning after a frost.

◁ On a freezing night, when water vapor meets a cold window pane, it leaves patterns on the glass.

▷ When water vapor freezes onto twigs and spiders' webs, tiny crystals of ice grow. This is frost.

Freezing and Thawing

In very cold weather the surface of the ground may freeze. If rain falls, it will freeze on the surface and form a layer of ice. Ice on the ground is very dangerous. It makes it difficult to walk or drive. Cars slide on the road. In icy weather water can freeze inside pipes. The pipes need to be wrapped with thick material or heated so that the water inside does not freeze.

▷ When the water in rocks freezes and then thaws, it makes rocks split. This is what has caused the shape of these rocks.

▽ Many car accidents in winter are caused by ice on the road.

◁ If ice melts from a rooftop during the day, a cold night may freeze the drips into icicles.

▷ Water can freeze in pipes and make them burst. Water will flood everywhere when the ice thaws.

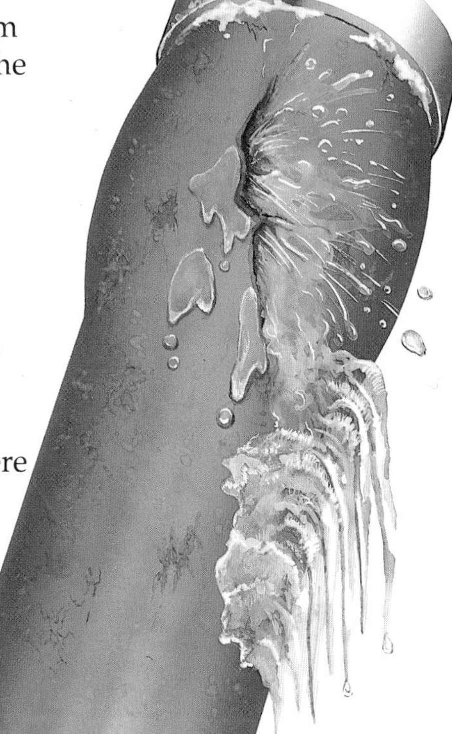

Blizzards

A heavy snowstorm with a wind blowing too, is called a **blizzard**. Blizzards are dangerous because the temperature is freezing, and it is difficult to see and move in it. The wind blows snow into deep piles called **drifts**. Many people, such as farmers, have jobs in which it is important to know whether blizzards are on their way.

▷ During a blizzard the snow falls heavily like a white curtain.

▷ Drifting snow can bury farm animals. Farmers have to go and rescue the animals because they may die.

▷ Blizzards are dangerous for air travel. Planes may not be able to take off, and runways have to be cleared.

Dangers of Snow and Ice

In the mountains, great masses of snow sometimes slide down the steep slopes. The falling snow buries everything in its path. This is an **avalanche** and it is very dangerous. Snow also causes other dangers. Drifts can block roads and cover signposts and road markings. People may lose their way and become stranded.

▷ Avalanches may start when the temperature rises and the snow begins to melt.

▽ Helicopters are used to rescue people who are stranded in the snow.

△ Snow makes everything look quite different. When it covers roads and signposts it is easy to lose one's way.

Clearing Snow

Heavy snowfalls cause problems in cities and in the country. **Snowplows** are used to clear roads and keep them open to traffic. If snow is forecast, trucks may spread salt or sand on the roads. Salt helps to melt the snow. The trucks spread sand to help car tires grip the surface of the road.

▽ Snowplows can help clear the roads of snow.

▷ Clearing snow from sidewalks makes it easier for people to walk.

▽ Roads are cleared of heavy snow. Salt and sand are spread on the roads to help make them safer.

Snow and Ice on Land

Snow does not fall evenly around the world. Some places never have snow. Mountains often have snowy **peaks** all year round. This is because it is much colder on high ground. Ice and snow may wear away or change the shape of rocks. Freezing and thawing causes the rocks to crack or break. On mountains, rivers of ice called **glaciers** grind away at the rock and make deep valleys.

▷ The peak of this mountain is covered in snow. Plants grow on the lower slopes where it is much warmer.

▽ Glaciers are rivers of ice. They cut away at the mountain rock. Pieces of rock are carried along and left at the bottom of the slope.

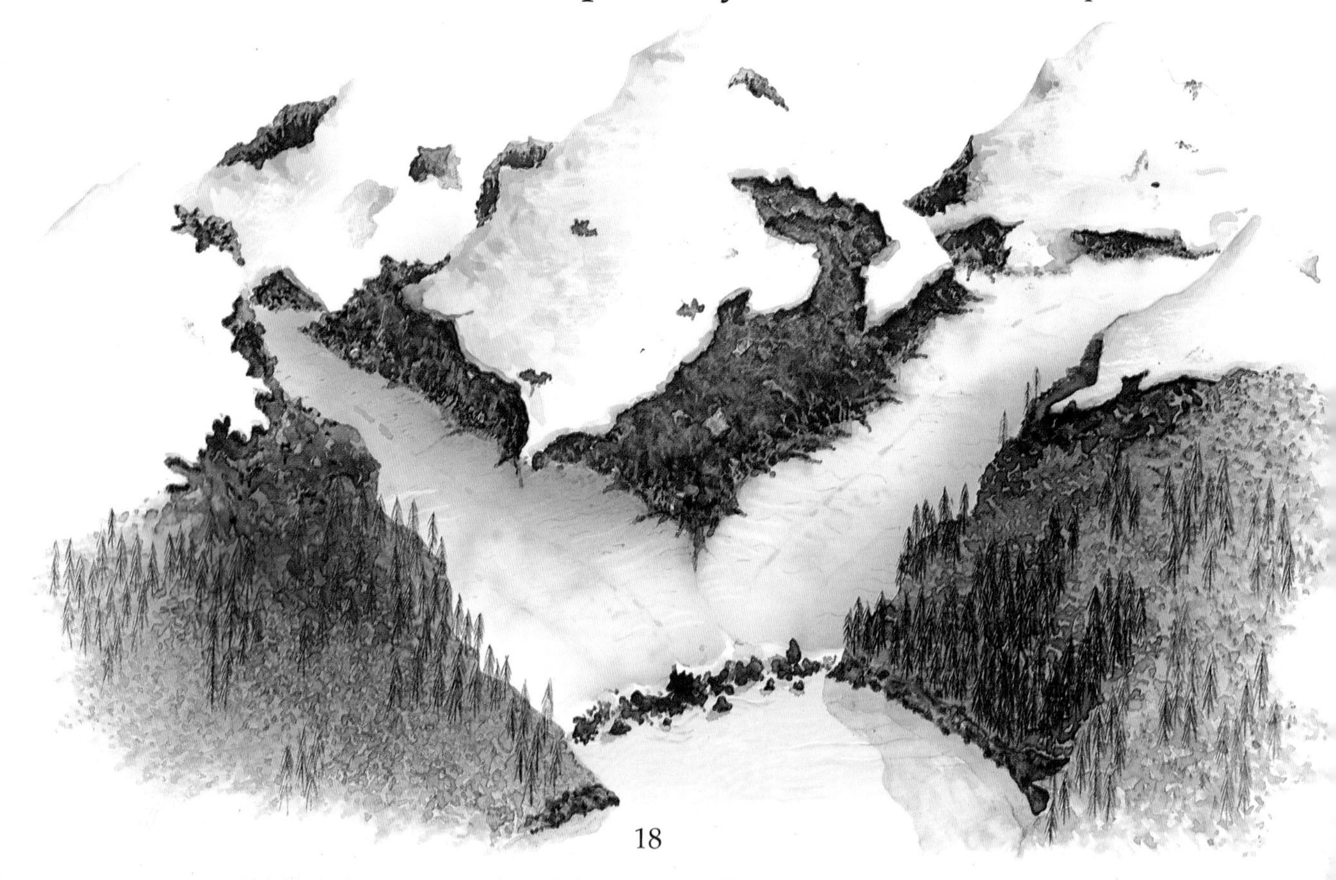

A Snowy World

The North and South Poles are at opposite ends of the Earth. The area around the North Pole is called the Arctic. It is mainly frozen ocean. The area around the South Pole is called the Antarctic. They are the coldest places on Earth. For much of the year the **polar lands** are frozen and the land is covered in snow and ice. In summer, some of the ice and snow thaws.

▽ In the Antarctic, almost all the land is covered with ice. Snow and ice are deeper in the Antarctic than anywhere else in the world.

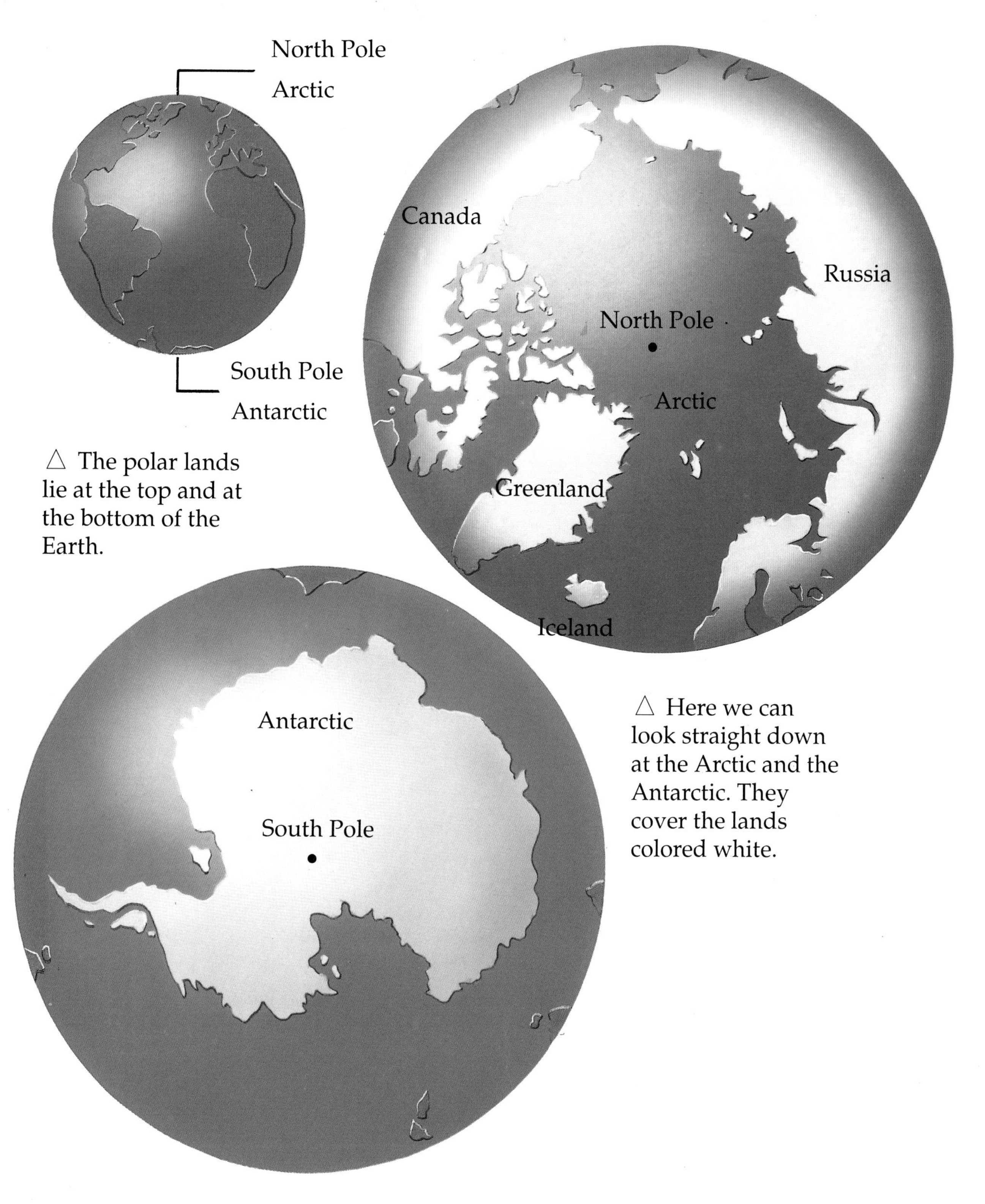

△ The polar lands lie at the top and at the bottom of the Earth.

△ Here we can look straight down at the Arctic and the Antarctic. They cover the lands colored white.

Living with Ice and Snow

People who live in the ice and snow have learned how to survive the harsh weather. Houses are built to keep in the warmth and not to fall down under the weight of the snow. People in the Arctic use snowmobiles to move around. They cross the ice more easily than vehicles with wheels. In other snowy places, people put chains on car wheels.

▽ In the Arctic, snowmobiles are used to get around.

▷ Houses in mountain areas have steep roofs to help the snow slide off.

▽ In places where it snows each winter, drivers put chains or special snow tires on their cars to help grip the road.

Plants of the Snow and Ice

Plants that grow in places where there is a lot of snow and ice have to be able to survive cold, wind, and little rain. Most plants that grow in polar lands or high on mountains are small and lie low on the ground. **Conifer** trees grow in many snowy places. These trees have springy branches which bend under the weight of the snow.

▷ Conifer trees have branches which bend to keep them from snapping under the heavy snow.

◁ Edelweiss grows in the mountains. It is covered with tiny hairs.

△ Crocus grow in early spring when there is still snow on the ground. They push through the snow to flower.

▷ Bearberry is a plant which grows in the Arctic. It is small and grows in clusters.

Animals of Snow and Ice

Many animals cannot survive in cold and icy weather. The snow covers the plants they eat. Some birds **migrate**, or go to warmer places before the winter begins. Other animals **hibernate**, sleeping through the cold months. Animals in polar lands have thick coats to survive the cold. Some animals living in cold places turn white so that they cannot be easily seen against the snow.

▷ Polar bears have thick coats and large furry pads on their feet.

▽ The Arctic stoat and hare both turn white in winter. This hides them in the snow.

◁ Penguins have a thick layer of fat to keep them warm. A chick is protected by sitting on the parent's feet and being covered.

▷ Swallows cannot survive the cold winter weather. They migrate to warmer places.

▽ When the ground is frozen, food is hard to find. Many birds rely on berries to survive.

◁ Chipmunks hibernate during the winter months.

Enjoying Snow and Ice

Would you like to live in the Arctic or the Antarctic? Many people dislike snow and ice because they cause so many problems.

If you dress warmly and are able to get around easily, snow can be a lot of fun. You can build things with the snow, and play all kinds of exciting games and sports.

▽ Cross-country skiing is popular in countries such as Norway, where there is plenty of snow in winter.

▽ Toboggans slide easily across the snow. They travel downhill fast.

△ Snow figures are fun to build, but they melt when the snow thaws.

◁ Ice hockey is a popular game in countries where it is very cold in the winter.

Things to Do

- Try catching a snowflake. You will have to wait for it to snow before you can do this! Ask an adult if you can borrow a hand lens. Put a pair of gloves or mittens in the freezer. Dress warmly to go outside. Take the cold gloves out of the freezer and put them on. Catch snowflakes on your gloves and quickly look at the snowflake with the hand lens. Do all the snowflakes have the same pattern?

- When the earth becomes hard and icy, birds find it difficult to find food. Put some seeds, nuts, peanut butter, or beef suet outside in the garden or on a window ledge. Watch how many different types of birds come to feed. You can make lists of which birds come each day.

Glossary

avalanche A large mass of loose snow that falls downhill.

blizzard A snowstorm where the wind is also blowing.

conifer A tree that has cones and stays green all year round.

drifts Deep banks of snow that the wind has piled up.

freezing point The temperature at which a liquid or gas freezes. The freezing point of water is 32°F.

glacier A river of ice that moves slowly down a mountain or across land.

hibernate To spend the winter in a deep sleep-like state.

ice crystals Structures that are patterned and made from ice. They form as tiny water droplets freeze.

migrate To travel regularly from one part of the world to another in certain seasons of the year. Swallows migrate to cool countries in summer and to warm countries in winter.

peak The top of a mountain. Some peaks are pointed.

polar lands The area of land and sea around the North and South Poles.

snowplow A bulldozer-like vehicle that is fitted with a piece of equipment to clear snow out of the way.

temperature The measurement of heat or coldness in a place, animals, or objects.

water cycle The endless pattern in which water, in the shape of rain, snow, or hail falls, turns into a gas and rises, turning into droplets of water again and falls once more.

water vapor The invisible gas in the air. Water becomes a gas when it evaporates.

Index

Photographic credits: B & C Alexander cover, 17, Bruce Coleman Ltd. (B. Coleman) 7, (F. de Nooyer) 8, (F. Eriza) 20, (T. Buchhol) 28; Robert Harding 11, 19, (R. Mcleod) 4; Hutchison Library (B. Regent) 3; © Julie Marcotte/Stock Boston 29; ZEFA cover, 13, 15, 23, 25, 27.